Uranium

Dr. Hemant Pathak

Copyright © 2017 Dr. Hemant Pathak

All rights reserved.

ISBN: 1976235057
ISBN-13: 978-1976235054

DEDICATION

Dedicated to Shri Sainath Maharaj the all omnipotent of world the most merciful.

CONTENTS

Foreword vi

Glossary ix

1	Introduction	13
2	Discovery of uranium	15
3	Occurrence	16
4	Isotopes	18
5	Unique property of Uranium	19
6	Uranium enrichment	20
7	Nuclear reactor	22
8	Nuclear power stations	23
9	Uranium ore and Mining	25
10	Uses of Uranium	26
11	Breeder reactor	27
12	Uranium as nuclear weapon	27
13	Uranium Resources and reserves	29
14	Human exposure	30
15	Effects and precautions	31
10	References	32

Foreword

Uranium is a very important element because it provides us nuclear fuel used to generate electricity in nuclear power stations. It is major material from which other synthetic transuranium elements are made.

Uranium was first discovered by Martin Klaproth in 1789. Uranium has been used in nuclear weapons, On August 6, 1945 the U.S. Bomber Enola Gay flew over the Japanese city of Hiroshima. Seconds later a metallic projectile fell towards its target. **Millions of people are being exposed to harmful levels of radioactive radiation.** In a blinding flash the world felt the power of a new age, the nuclear age.

It also used in the making of special chemicals and electronic parts. small amounts of uranium are used to produce radioactive isotopes that are used in the medical, industrial, and research fields.

This book provides an basic guide to researchers, it offers: A key point study of the valuable metal and its impact on current Society. Simply explained, Uranium is an important book bringing together diverse viewpoints from Scientist, state agencies and regulators, for all who wish to save Earth with quality Life.

Dr. Hemant Pathak

M.Sc. (Gold medalist), Ph. D.

Assistant Professor of Engineering Chemistry

Indira Gandhi Govt. Engineering

college, Sagar, MP, India

Acronyms and Symbol

Becquerel: SI unit of intrinsic radioactivity in a material. one Bq indicates one disintegration per second. The activity of a quantity of radioactive material which averages one decay per second.

Giga: One Billion Units (10^9 Watts).

Gray (GY): The SI unit of absorbed radiation dose, one joule per kilogram of tissue.

MegaWatt (MW): A unit of power (10^6 watts).

Tera: One million million units (one billion kWh).

Glossary

Actinide: An element with atomic number of 89 (actinium) to 103. (also called transuranics). Actinides are radioactive and typically have long half-lives. They are therefore significant in wastes arising from nuclear fission, e.g. used fuel.

Activation product: A radioactive isotope of an element, which has been created by neutron bombardment.

Activity: The number of disintegrations per unit time inside a radioactive source. Expressed in becquerels.

Alpha particle: A positively-charged particle emitted from the nucleus of an atom during radioactive decay. Alpha particles are helium nuclei, with 2 protons and 2 neutrons.

Atom: A particle of matter which cannot be broken up by chemical means. Atoms have a nucleus consisting of positively-charged protons and uncharged neutrons of almost the same mass.

Beta particle: A particle emitted from an atom during radioactive decay. Beta particles are generally electrons but may be positrons.

Biological shield: A mass of absorbing material placed around a reactor or radioactive material to reduce the radiation to a level safe for humans.

Centrifuge: A cylinder spinning at high speed to physically separate gas components of slightly different mass, e.g. uranium hexafluoride with U-235 and U-238 atoms.

Chain reaction: A reaction that stimulates its own repetition, in particular where the neutrons originating from nuclear fission cause an ongoing series of fission reactions.

Control rods: Devices to absorb neutrons so that the chain reaction in a reactor core may be slowed or stopped by inserting them further, or accelerated by withdrawing them.

Conversion: Chemical process turning U_3O_8 into UF_6 preparatory to enrichment.

Coolant: The liquid or gas used to transfer heat from the reactor core to the steam generators or directly to the turbines.

Core: The central part of a nuclear reactor containing the fuel elements and any moderator.

Critical mass: The smallest mass of fissile material that will support a self-sustaining chain reaction under specified conditions.

Decay: Disintegration of atomic nuclei resulting in the emission of alpha or beta particles (usually with gamma radiation). Also the exponential decrease in radioactivity of a material as nuclear disintegrations take place and more stable nuclei are formed.

Deuterium: Heavy hydrogen, a stable isotope having one proton and one neutron in the nucleus. It occurs in nature as 1 atom to 6500 atoms of normal hydrogen.

Disintegration: natural change in the nucleus of a radioactive isotope as particles are emitted, making it a different element.

Element: A chemical substance that cannot be divided into simpler substances by chemical means; atomic species with same number of protons.

Enriched uranium: Uranium in which the proportion of U-235 has been increased above the natural 0.7%. Reactor-grade uranium is usually enriched to about 3.5% U-235, weapons-grade uranium is more than 90% U-235.

Enrichment: Physical process of increasing the proportion of U-235 to U-238.

Fast breeder reactor (FBR): A fast neutron reactor configured to produce more fissile material than it consumes, using fertile material such as depleted uranium in a blanket around the core.

Fast neutron: neutron released during fission, travelling at very high velocity (20,000 km/s) and having high energy.

Fission: The splitting of a heavy nucleus into two, accompanied by the release of a relatively large amount of energy and usually one or more neutrons. It may be spontaneous but usually is due to a nucleus absorbing a neutron and thus becoming unstable.

Gamma rays: High energy electro-magnetic radiation from the atomic nucleus, virtually identical to X-rays.

Genetic mutation: Sudden change in the chromosomal DNA of an individual gene. It may produce inherited changes in descendants. Mutation in some organisms can be made more frequent by irradiation.

Half-life: The period required for half of the atoms of a particular radioactive isotope to decay and become an isotope of another element.

Heavy water: Water containing an elevated concentration of molecules with deuterium atoms.

Isotope: An atomic form of an element having a particular number of neutrons. Different isotopes of an element have the same number of protons but different numbers of neutrons and hence different atomic masses, eg. U-235, U-238.

Natural uranium: Uranium with an isotopic composition as found in nature, containing 99.3% U-238, 0.7% U-235 and a trace of U-234. It can be used as fuel in heavy water-moderated or graphite-moderated reactors.

Neutron: An uncharged elementary particle found in the nucleus of every atom except hydrogen. Solitary mobile neutrons travelling at various speeds originate from fission reactions. Slow neutrons can in turn readily cause fission in nuclei of fissile isotopes, e.g. U-235, Pu-239, U-233; and fast neutrons can cause fission in nuclei of fertile isotopes such as U-238, Pu-239. Sometimes atomic nuclei simply capture neutrons.

Nuclear reactor: A device in which a nuclear fission chain reaction occurs under controlled conditions so that the heat yield can be harnessed or the neutron beams utilised. All commercial reactors are thermal reactors, using a moderator to slow down the neutrons.

Nuclide: elemental matter made up of atoms with identical nuclei, therefore with the same atomic number and the same mass number.

Radiation: The emission and propagation of energy by means of electromagnetic waves or particles.

Radioactivity: The spontaneous decay of an unstable atomic nucleus, giving rise to the emission of radiation.

Radionuclide: A radioactive isotope of an element.

Radiotoxicity: The adverse health effect of a radionuclide due to its radioactivity.

Uranium (U): A mildly radioactive element with two isotopes which are fissile (U-235 and U-233) and two which are fertile (U-238 and U-234). Uranium is the basic fuel of nuclear energy.

Uranium hexafluoride (UF_6): A compound of uranium which is a gas above 56°C and is thus a suitable form in which to enrich the uranium.

1. Introduction

Uranium (symbol **U**) is a chemical element. It is a silvery-white metal in the actinide series of the periodic table. Uranium has a melting point of 1132°C.

- Atomic number: 92
- Atomic weight (average mass of the atom): 238.02891
- Density: 18.95 grams per cubic centimeter
- Phase at room temperature: Solid
- Melting point: 2,075 degrees Fahrenheit
- Boiling point: 7,468 F (4,131 C)
- Number of isotopes: 16, 3 naturally occurring
- Most common isotopes:
    1. U-234 (0.0054 percent natural abundance),
    2. U-235 (0.7204 percent natural abundance),
    3. U-238 (99.2742 percent natural abundance)
- Uranium's most stable isotope, uranium-238, has a half-life of about 4,468,000,000 years. It decays into thorium-234 through alpha decay or decays through spontaneous fission.
- Estimated Crustal Abundance: 2.7 milligrams per kilogram
- Estimated Oceanic Abundance: 3.2×10^{-3} milligrams per liter

- Number of Stable Isotopes: **0**

- Ionization Energy: 6.194 eV

- Oxidation States: +6, +5, +4, +3

- Electron Shell Configuration:

$1s^2$
$2s^2\ 2p^6$
$3s^2\ 3p^6\ 3d^{10}$
$4s^2\ 4p^6\ 4d^{10}\ 4f^{14}$
$5s^2\ 5p^6\ 5d^{10}\ 5f^3$
$6s^2\ 6p^6\ 6d^1$
$7s^2$

Uranium is a silvery white, weakly radioactive metal. It has 92 protons and 92 electrons, of which 6 are valence electrons. It is weakly radioactive because all isotopes of uranium are unstable, with half-lives varying between 159,200 years and 4.5 billion years. It has a Mohs hardness of 6, sufficient to scratch glass and approximately equal to that of titanium, rhodium, manganese and niobium. It is malleable, ductile, slightly paramagnetic, strongly electropositive and a poor electrical conductor. Uranium metal has a very high density of 19.1 g/cm³, denser than lead (11.3 g/cm³), but slightly less dense than tungsten and gold (19.3 g/cm³).

Finely divided uranium metal presents a fire hazard because uranium is pyrophoric; small grains will ignite spontaneously in air at room temperature. Uranium metal is commonly handled with gloves as a sufficient precaution. Uranium concentrate is handled and contained so as to ensure that people do not inhale or ingest it.

2. Discovery of uranium

Uranium was discovered in 1789 by Martin Klaproth, a German chemist, in the mineral called pitchblende, who named the new element after the planet Uranus. which had been discovered eight years earlier. Uranium is 18.7 times as dense as water.

Eugene-Melchior Peligot was the first person to isolate the metal.

Uranium was found to be radioactive in 1896 by Antoine H. Becquerel, a French physicist. Becquerel had left a sample of uranium on top of an unexposed photographic plate, which became cloudy. He concluded it was giving off invisible rays.

This was the first instance that radioactivity had been studied and opened up a new field of science.

Marie Curie with Pierre Curie (Polish scientist) coined the term radioactivity shortly after Becquerel's discovery, continued the research to discover other radioactive elements, such as polonium and radium, and their properties.

The human body contains approximately 56 μg of uranium, 32 μg (56%) are in the skeleton, 11 μg in muscle tissue, 9 μg in fat, 2 μg in blood and less than 1 μg in lung, liver and kidneys.

3. Occurrence

Uranium is a very heavy metal which has been used as an abundant source of concentrated energy for 62 years. It occurs naturally in low concentrations of a few parts per million in soil, rock and water, and is commercially extracted from uranium-bearing minerals such as uraninite.

Uranium occurs in most rocks in concentrations of 2 to 4 parts per million and is as common in the Earth's crust as other metals.

The daily intake of uranium is estimated to be 1–2 μg in food and 1.5 μg in water consumed. It also occurs in seawater, and can be recovered from the oceans. It is 48th among the most abundant elements found in natural crustal rock, 40 times more abundant than silver.

Due to its presence in soil, rocks, surface and underground water, air, plants, and animals it occurs also in trace amounts in many foods and in drinking water.

Though uranium is highly associated with radioactivity, its rate of decay is so low that this element is actually not one of the more radioactive ones out there.

Uranium-238 has a half-life of an incredible 4.5 billion years. Uranium-235 has a half-life of over 700 million years. Uranium-234 has the shortest half-life at 245,500 years, but it occurs only indirectly from the decay of U-238.

Uranium is the only naturally formed in supernovae. Uranium is the heaviest naturally occurring element and is found at an average concentration of 3 mg/kg in the earth's crust. In seawater the concentration is about 3.0 µg/l.

Naturally occurring radioactive uranium found usually in the form of uranium dioxide (UO_2), is most commonly used in the nuclear power industry to generate electricity.

The uranium in the human body is derived mostly from uranium in food, especially from vegetables, cereals, and table salt.

4. Isotopes

Uranium occurs in several slightly differing forms known as isotopes. These isotopes differ from each other in the number of uncharged particles (neutrons) in the nucleus.

Naturally occurring uranium consists of three isotopes: uranium-234 (p-92,n-142) uranium-235(p-92,n-143) and uranium-238 (p-92,n-146). It is found as uranium-238 (99.27%), uranium-235 (0.72%), and a very small amount of uranium-234 (0.0054%).

Although all three isotopes are radioactive, only uranium-235 is a fissionable material so it can be used for nuclear power.

Uranium decays slowly by emitting an alpha particle. The half-life of uranium-238 is about 4.47 billion years and that of uranium-235 is 704 million years, making them useful in dating the age of the Earth.

Isotope	Atomic mass	Natural abundance (%)	Half life	Mode of decay
^{233}U	233.040	-	1.590×10^5 y	α
			$> 2.7 \times 10^{17}$ y	sf
^{234}U	234.041	0.0054	2.453×10^5 y	α
			1.5×10^{16} y	sf
^{235}U	235.044	0.7204	7.03×10^8 y	α
			1.0×10^{19} y	sf
^{236}U	236.046	-	2.342×10^7 y	α
			2.5×10^{16} y	sf

^{238}U	238.051	99.2742	8.2×10^{15} y	sf

5. Unique property of Uranium

Uses of Uranium exploit its unique nuclear properties. Uranium-235 is the only naturally occurring fissile isotope, which makes it widely used in nuclear power plants and nuclear weapons.

High density uranium uses in the keels of yachts and as counterweights for aircraft control surfaces, as well as for radiation shielding.

Physicists Otto Hahn and Fritz Strassman (1938) discovered something unique about uranium, When a fissionable material is struck by a neutron, its nucleus can release energy by splitting into smaller fragments. it would split into two nearly equal parts—a process called nuclear fission. If some of the fragments are other neutrons, they can strike other atoms and cause them to split as well. A fissionable material, such as uranium-235, is a material capable of producing enough free neutrons to sustain a nuclear chain reaction.

If those neutrons go on to split other uranium atoms, creates a fission chain reaction and when this happens millions of times, it can create a lot of heat from relatively small amounts of uranium.

6. Uranium enrichment

Civilian and military reactors require uranium that has a higher proportion of uranium-235 than present in natural uranium.

The process used to increase the amount of uranium-235 relative to uranium-238 is known as uranium enrichment.

U-238 decays very slowly, its half-life being about the same as the age of the Earth (4500 million years). This means that it is barely radioactive, less so than many other isotopes in rocks and sand. Nevertheless it generates 0.1 watts/tonne as decay heat and this is enough to warm the Earth's core. U-235 decays slightly faster.

Uranium needs to undergo enrichment so that enough uranium-235 is present. Uranium-235 and to a lesser degree uranium-233 have a much higher fission cross-section for slow neutrons.

Uranium-233, can be produced from natural thorium and is also important in nuclear technology. Uranium-238 has a small probability for spontaneous fission or even induced fission with fast neutrons.

Uranium is generally used in reactors in the form of uranium dioxide (UO_2), nuclear weapons use the metallic form. Production of uranium dioxide or metal requires chemical processing of yellowcake.

For most of the world's reactors, the next step in making the fuel is to convert the uranium oxide into a gas, uranium hexafluoride (UF_6), which enables it to be enriched. A major hazard in both the uranium conversion and uranium enrichment processes comes from uranium hexafluoride, which is chemically toxic as well as radioactive. After enrichment, the UF_6 gas is converted to uranium dioxide (UO_2), which is formed into fuel pellets. These fuel pellets are placed inside thin metal tubes, then known as fuel rods, which are assembled in bundles to become the fuel elements or assemblies for the core of the reactor. Conversion and enrichment facilities have had a number of accidents involving uranium hexafluoride.

The bulk of waste from the enrichment process is depleted uranium. Most of the uranium-235 has been extracted from it. Depleted uranium has been used to fabricate armor-piercing conventional weapons and tank armor plating.

Enrichment increases the proportion of the uranium-235 isotope from its natural level of 0.7% to 4 - 5%. This enables greater technical efficiency in reactor design and operation, particularly in larger reactors, and allows the use of ordinary water as a moderator.

In a typical large power reactor there might be 51,000 fuel rods with over 18 million pellets.

7. Nuclear reactor

The isotope U-235 is important because under certain conditions it can readily be split, yielding a lot of energy. It is said to be fissile, its nucleus can be split by thermal neutrons, neutrons with the same energy as their ambient surroundings. The nucleus of a U-235 atom has 143 neutrons.

When a free neutron bumps into the atom, it splits the nucleus, throwing off additional neurons, which can then zing into the nuclei of nearby U-235 atoms, creating a self-sustaining cascade of nuclear fission.

Uranium-238 is fissionable by fast neutrons, and is fertile, meaning it can be transmuted to fissile plutonium-239 in a nuclear reactor.

This fission consequently generate heat. In a, this heat is used to boil water, creating steam that turns a turbine to generate power, and the reaction is controlled by materials such as cadmium or boron, which can absorb extra neutrons to take them out of the reaction chain.

Chain reaction over and over again, many millions of times, a very large amount of heat is produced from a relatively small amount of uranium. Burning uranium occurs in a nuclear reactor. The heat is used to make steam to produce electricity.

In reactor natural uranium as their fuel, graphite or heavy water as a moderator, U_3O_8 concentrate simply needs to be refined and converted directly to uranium dioxide.

The enrichment process can also be reversed. Highly enriched uranium can be diluted, or "blended down" with depleted, natural, or very low-enriched uranium to produce 3 to 5 percent low-enriched reactor fuel. Uranium metal at various enrichments must be chemically processed so that it can be blended into a homogeneous material at one enrichment level. As a result, the health and environmental risks of blending are similar to those for uranium conversion and enrichment.

The fuel elements are surrounded by a substance called a moderator to slow the speed of the emitted neutrons and thus enable the chain reaction to continue. Water, graphite and heavy water are used as moderators in different types of reactors.

8. Nuclear power stations

One kilogram of uranium-235 can theoretically produce about 20 terajoules of energy (2×10^{13} joules), assuming complete fission; as much energy as 1500 tonnes of coal.

Commercial nuclear power plants use fuel that is typically enriched to around 3% uranium-235. 440 nuclear reactors with a total output capacity of about 390,000 megawatts (MW) operating in 31 countries. Over 60 more reactors are under construction and another 160 are planned.

Bulgaria, Czech Republic, Finland, France, Hungary, South Korea, Slovakia, Slovenia, Belgium, Switzerland and Ukraine all get 30% or more of their electricity from nuclear reactors.

Due to disarmament (after 1990) a lot of military uranium has become available for electricity production. The military uranium is diluted about 25:1 with depleted uranium from the enrichment process before being used in power generation.

The USA has about one hundred reactors operating, supplying 20% of its electricity. France gets three quarters of its electricity from uranium.

Over the 60 years that the world has enjoyed the benefits of cleanly-generated electricity from nuclear power, there have been 17,000 reactor-years of operational experience.

In case of Nuclear power stations require heat to produce steam to drive turbines and generators. In a nuclear power station, however, the fissioning of uranium atoms replaces the burning of coal or gas. In a nuclear reactor the uranium fuel is assembled in such a way that a controlled fission chain reaction can be achieved. The heat created by splitting the U-235 atoms is then used to make steam which spins a turbine to drive a generator, producing electricity.

The chain reaction is carefully controlled using neutron-absorbing materials. The heat generated by the fuel is used to create steam to turn turbines and generate electrical power.

Because of the kind of fuel used (U-235), if there is a major uncorrected malfunction in a reactor the fuel may overheat and melt, but it cannot explode like a bomb.

A typical 1000 megawatt (MWe) reactor can provide enough electricity for a modern city of up to one million people.

9. **Uranium ore and Mining**

Uranium is widespread in many rocks, and even in seawater. It is seldom sufficiently concentrated to be economically recoverable. Uranium is sold only to countries which are signatories of the Nuclear Non-Proliferation Treaty, and which allow international inspection to verify that it is used only for peaceful purposes.

Uranium ore can be mined by underground or open-cut methods, depending on its depth. After mining, the ore is crushed and ground up. Then it is treated with acid to dissolve the uranium, which is recovered from solution.

Uranium may also be mined by in situ leaching, where it is dissolved from a porous underground ore body in situ and pumped to the surface.

When the uranium fuel has been in the reactor for about three years, the used fuel is removed, stored, and then either reprocessed or disposed of underground.

An approximate 4.6 billion tonnes of uranium are estimated to be in sea water. Bakouma in the prefecture of Mbomou in Central African Republic have significant reserve of uranium.

10. Uses of Uranium

Uranium is a very important element because it provides us with nuclear fuel used to generate electricity in nuclear power stations. It is also the major material from which other synthetic transuranium elements are made.

Uranium compounds have been used for centuries to color glass. A 2,000 year old sample of yellow glass found near Naples, Italy contains uranium oxide.

Uranium trioxide (UO_3) used in the manufacture of Fiestaware plates. Other uranium compounds have also been used to make vaseline glass and glazes. The uranium within these items is radioactive and should be treated with care.

Depleted uranium is uranium that has much less uranium-235 than natural uranium. **Depleted uranium is used in kinetic energy penetrators and armor plating.** It is considerably less radioactive than natural uranium. It is a dense metal that can be used as ballast for ships and counterweights for aircraft. It is also used in ammunition and armour. This ammunition consists of depleted uranium alloyed with 1–2% titanium or molybdenum.

Uranium is used as a colorant in uranium glass, producing lemon yellow to green colors. Uranium glass fluoresces green in ultraviolet light. It was also used for tinting and shading in early photography.

At high impact speed, the density, hardness, and pyrophoricity of the projectile enable the destruction of heavily armored targets. Tank armor and other removable vehicle armor can also be hardened with depleted uranium plates.

Uranium is also used by the military to power nuclear submarines and in nuclear weapons.

11. Breeder reactor

In a breeder reactor uranium-238 captures neutrons and undergoes negative beta decay to become plutonium-239. This synthetic, fissionable element can also sustain a chain reaction.

12. Uranium as nuclear weapon

Otto Hahn, Enrico Fermi and J. R. Oppenheimer starting, its use as a fuel in the nuclear power industry and in the first nuclear weapon named as Little Boy used in war. Only 1.38 percent of the uranium in the Little Boy bomb that destroyed Hiroshima underwent fission. The bomb contained about 64 kg of uranium total amount.

Arms race during the cold war between the United States and the Soviet Union produced tens of thousands of nuclear weapons that used uranium metal and plutonium-239.

This waste product was diverted to the glazing industry, making uranium glazes very inexpensive and abundant. Besides the pottery glazes, uranium tile glazes accounted for the bulk of the use, including common bathroom and kitchen tiles which can be produced in green, yellow, mauve, black, blue, red and other colors.

Uranium was also used in photographic chemicals, in lamp filaments for stage lighting bulbs, to improve the appearance of dentures, and in the leather and wood industries for stains and dyes. Uranium salts are mordants of silk or wool.

Uranyl acetate and uranyl formate are used as electron-dense stains in transmission electron microscopy viz. staining of viruses, isolated cell organelles and macromolecules.

The discovery of the radioactivity of uranium ushered in additional scientific and practical uses of the element. The long half-life of the isotope uranium-238 (4.51×10^9 years) makes it well-suited for use in estimating the age of the earliest igneous rocks and for other types of radiometric dating, including uranium-thorium dating, uranium-lead dating and uranium-uranium dating. Uranium metal is used for X-ray targets in the making of high-energy X-rays. The Little Boy bomb detonated 509 meters above Hiroshima and left only the frames of a few reinforced concrete buildings standing in the mile radius around Ground Zero, all humans and animals are naturally exposed to minute amounts of uranium from food, water, soil and air.

13. Uranium Resources and reserves

Uranium is mined in 20 countries, with over half coming from Canada, Kazakhstan, Australia, Niger, Russia and Namibia.

Australia has 31% of the world's known uranium ore reserves and the world's largest uranium deposit, located at the **Olympic Dam Mine**. It is estimated that 5.5 million tonnes of uranium exists in ore reserves, while 35 million tonnes are classed as mineral resources.

Kazakhstan's are 13%, Canada's and Russia's each 9%. Kazakhstan's are over 700,000 tonnes of uranium and Canada's and Russia's are over 500,000 tonnes of uranium.

Several countries have significant uranium resources. Apart from the top four, they are in order: South Africa, Niger, Brazil, China, Namibia, Mongolia, Uzbekistan, and Ukraine, all with 2% or more of world total. Other countries have smaller deposits which could be mined if needed.

Kazakhstan is the world's top uranium producer, followed by Canada and then Australia as the main suppliers of uranium to world markets.

14. Human exposure

Accidental inhalation exposure to a high concentration of uranium hexafluoride has resulted in human fatalities, those deaths were associated with the generation of highly toxic hydrofluoric acid and uranyl fluoride rather than with uranium itself.

Human exposed to uranium by inhaling dust in air or by contaminated water and food. Radiation exposure and an increased risk for cancer is generally quite long the increase of cancer risk due to low radiation doses by linear extrapolation from highly exposed human populations, such as the survivors of Hiroshima and Nagasaki. The amount of uranium in air is usually very small. Radiological effects are generally local because alpha radiation, the primary form of ^{238}U decay, has a very short range, and will not penetrate skin. Labour who work in factories that process phosphate fertilizers, live near nuclear test range or live or work near a coal-fired power plant, may have increased exposure to uranium.

Uranyl ions, which accumulate in bone, kidney, liver, and reproductive tissues. Uranium can be decontaminated from steel surfaces and aquifers.

15. Effects and precautions

Uranium is a weak radioactive toxic metal. Human body organs like kidney, brain, liver, heart, and other systems can be affected by uranium exposure behave as reproductive toxicant. Alpha radiation from inhaled uranium has been demonstrated to cause lung cancer in exposed nuclear workers.

Uranyl ions, such as from uranium trioxide or uranyl nitrate and other hexavalent uranium compounds, have been shown to cause birth defects and immune system damage in laboratory animals and its decay products. Exposure to strontium-90, iodine-131, and other fission products is unrelated to uranium exposure, but may result from medical procedures or exposure to spent reactor fuel or fallout from nuclear weapons.

16. References

1. Anderson, M., Enyeart, T.D., Jackson, T.L., Smith, R.W., Stewart, R.A., Thomson, R.A., Ulick, M.D., Zander, K.K., 1997. Resumption of Use of Depleted Uranium Rounds at Nellis Air Force Range. USAir Force, Target 63-10.

2. ATSDR (Agency for Toxic Substances and Disease Registry), 1999. US Public Health Service, Department of Health & Human Services. Toxicological Profile for Uranium, Atlanta, GA.

3. Bellis, D., Ma, R., Bramall, N., McLeod, C.W., 2001a. Airborne emission of enriched uranium at Tokaimura, Japan. The Science of the Total Environment 264, 283–286.

4. A. Bleise, P.R. Danesi, W. Burkart, Properties, use and health effects of depleted uranium (DU): a general overview, Journal of Environmental Radioactivity 64 (2003) 93–112.

ABOUT THE AUTHOR

Dr. Hemant Pathak held positions as Assistant Professor in the department of chemistry, Govt. Indira Gandhi Engineering College, Sagar, MP, India. He had extensive experience in teaching, research and administrative management.

Dr. Pathak received his Ph.D. degree in chemistry from Dr. Hari Singh Gour Central University, Sagar, India and M.Sc. Gold medalist from Jiwaji University, Gwalior. He has published 25 books and more than 50 research papers in reputed International and National journals and received several awards. He is a member of editorial boards and reviewer boards of several international journals and societies. His area of specialization includes Engineering Chemistry, Energy audits and Environmental Pollution management.

www.ingramcontent.com/pod-product-compliance
Lightning Source LLC
Chambersburg PA
CBHW050035230526

45470CB00003B/1300